Elasticity

© sciencebod 2016

Preface

Material properties like elasticity are usually studied to be able to select best materials for particular applications. Knowledge of elasticity is usually applied in almost all mechanical designs and structural analysis. The underlying idea in most applications is that materials should not be subjected to stress beyond the elastic limit, else they will get deformed.
This book presents the fundamental ideas on elasticity in very easy-to-understand manner. Ideas discussed include Hooke's law, tensile stress, tensile strain, Young's modulus, energy stored in elastic materials, etc. There are lots of numerical examples and exercises that have been used to help facilitate the students' quantitative understanding of the subject. So, read and enjoy!

© sciencebod 2016; okodan2003@gmail.com; +2348136094616, +2347035272976

Elasticity

Introduction

1

When a catapult is stretched, it quickly snaps back to its former shape once it is released. A wooden or plastic mater rule when bent will return to its former shape when the bending force is removed. When some load is suspended from the end of a spring of wire, the spring stretches a certain length but immediately the load is removed, the spring returns to its former length.

Such objects which regain their shape and size after the force causing the change is removed are said to be elastic objects and this properly (or phenomenon) is called elasticity.

Definition

2

A body is said to be elastic if it regains its original shape and size after undergoing a stretching or compression.
Elasticity is therefore, the ability of a substance to regain its original shape and size after being distorted by an external force.

Elastic limit of elastic materials

3

When stretching a catapult, the extension produced increases as the applied force increases until a particular point when further extension will lead to a distortion of the rubber of the catapult. At this point the force becomes too much for the rubber to bear and the extension will no longer be commensurate to the applied force. The rubber is then said to have reached its <u>elastic limit</u>.

Hooke's Law

4

Hooke's law states that provided the elastic limit of an elastic material is not exceeded, the extension, e, produced on the material is directly proportional to the load or applied force F.

That is:
$F \propto e$
or $F = ke$ -- (1)
where k is a constant of proportionality called elastic constant or force constant or stiffness of the material.

If F is Newtons and e is in meters, the elastic constant k will be in Nm^{-1}:

$$k = \frac{F}{e} \quad (Nm^{-1})$$ -- (2)

By definition, k is the force required to produce an extension of 1 meter on the material.

The knowledge of Hooke's law has been applied to developing a lot of equipments, one of which is the spring balance (used to measure the weight of objects). The spring balance is well discussed in our book on 'Measurement in Physics'.

Now, let's illustrate Hooke's law with the following JAMB question

5

A force of 15N stretches a spring to a total length of 30cm. An additional force of 10N stretches the spring 5cm further. Find the length of the spring

(A) 24cm   (B) 22.5cm   (C) 20.0cm   (D) 15.0cm

Solution

6

From the question,
The length of the spring plus the extension produced by the 15N force is:
$L + e = 30$ cm ----(i)
Where L = length of the spring, and e = extension produced.

But from Hooke's Law:
$F = ke$
Implying that $15 = ke$ ----(ii)

Next! An additional force of 10N stretches the spring 5cm further, implying that the extension is 5cm (or 0.05m) when the applied force is 10N. Substituting into Hooke's law gives:

$10 = k \times 0.05$
$\Rightarrow k = \dfrac{10}{0.05} = 200 \text{Nm}^{-1}$

Now, putting this value of k into equation (ii) gives:
$15 = 200e$
$\Rightarrow e = \dfrac{15}{200} = 0.075 \text{m} = 7.5 \text{cm}$

Finally, we get the length of the spring by putting this value of e into equation (i), this gives:

$L + 7.5 \text{cm} = 30 \text{cm}$
$\Rightarrow L = 30 \text{cm} - 7.5 \text{cm} = 22.5 \text{cm}$

So, option B is correct.

Next Illustration!

7

The spiral spring of a balance is 25.0cm long when 5N hangs on it and 30.0cm long when the weight is 10N. What is the length of the spring if the weight is 3N assuming Hooke's law is obeyed?

(A) 15cm (B) 17cm (C) 20cm (D) 23cm   **[JAMB]**

Solution!

8

For the 5N force;   L+e = 25cm
$\Rightarrow$ e=25-L
Putting this into Hooke's law, F=ke, we get:
5=k(25-L) --- (iii)

Doing same for the 10N force, we get
10=k(30-L) -- (iv)

Dividing equation (iii) by (iv) gives

$$\frac{5}{10} = \frac{k(25-L)}{k(30-L)}$$

$$\frac{5}{10} = \frac{25-L}{30-L}$$

$\Rightarrow$ 5(30-L) = 10(25-L)

150 - 5L = 250 − 10L

10L − 5L = 250 − 150

5L = 100

And so, L=20cm

This is the length of the spring when there is no weight on it.

Next we need to get the value of the elastic constant, k, by substituting L=20cm in equation (iii):
5 = k (25 – 20)
$\Rightarrow$ 5 = 5k
$\therefore$ k = 1N/cm

And finally, we get the extension when a weight of 3N hangs on the spring using Hookes law:
F = ke
3 = 1e
$\Rightarrow$ e = 3cm

And since the original length is L = 20cm,
the final length when a 3N weight hangs on it is 20cm + 3cm = 23cm

This is another JAMB question for you to try on your own!

9

An elastic material has a length of 36cm when a load of 40N is hung on it and a length of 45cm when a load of 60N is hung on it. The original length of the material is

(A) 12cm   (B) 20cm   (C) 18cm   (D) 15cm

Do it on your own before looking here!

10

If you got 18cm, then you are right!
This is how!

For the 40N force;   L+e = 36cm
$\Rightarrow$ e=36-L

Putting this into Hooke's law, F=ke, we get:
40=k(36-L) -- (v)

Doing same for the 60N force, we get
60=k(45-L) -- (vi)

Dividing equation (v) by (vi) gives:
$$\frac{40}{60} = \frac{k(36-L)}{k(45-L)}$$

$$\frac{40}{60} = \frac{36-L}{45-L}$$

$$\frac{2}{3} = \frac{36-L}{45-L}$$

$\Rightarrow$ 2(45-L) = 3(36-L)

90 - 2L = 108 – 3L

3L – 2L = 108 – 90

$\Rightarrow$ L = 18cm

This is the length of the original length of the material.

An experiment to verify Hooke's law

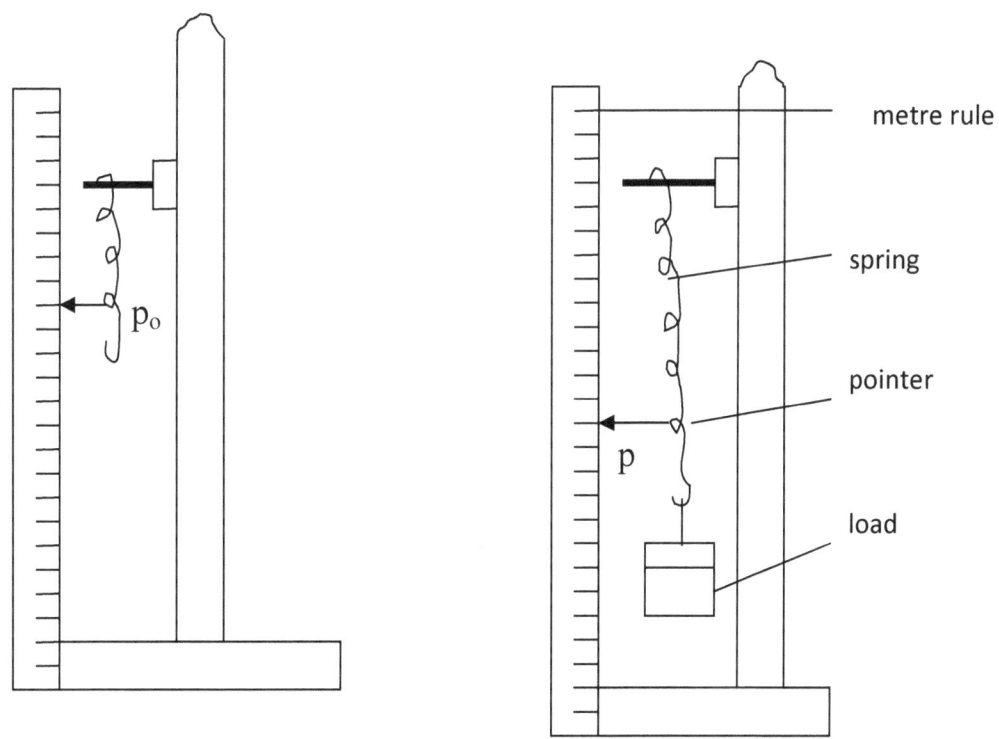

Figure 1. Experimental setup for verification of Hooke's law

The apparatus used in the verification of Hooke's law is shown in Figure 1. First, the pointer reading, p_o, is noted when no weight is suspended on the spring. Next, a load, w, is suspended on the spring and the new pointer reading, p, is noted. The extension, e, is then computed from the above readings as:
$e = p - p_o$

The experiment is repeated for fire more different values of w within the elastic limit of the spring. A graph of w (load) against e (extension) is plotted. This usually should give a straight line passing through the origin as shown in Plan 12. The graph shows that the load is directly proportional to the extension, and so verifies Hooke's law.

The graph load (w) against extension (e)

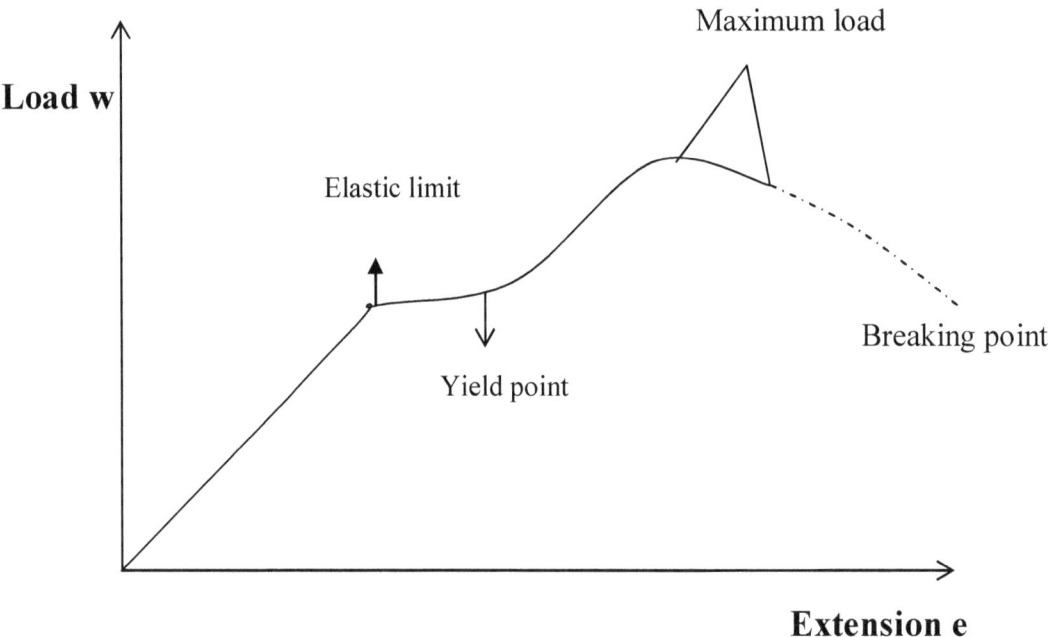

Figure 2. Graph of Load versus Extension for an elastic material

Figure 2 above shows the graph of load against extension for an elastic material which has be stretched gradually until it has exceeded its elastic limit.

The Elastic limit and Yield point

Hooke's law is obeyed until the elastic limit is attained. For loads beyond this point, the wire stretches permanently and if the load is removed, the wire is found to have been distorted and does not return to its original length.

Further increase in load results to a point where a small increase in load produces a large extension. This point is called the yield point. The yield point is the point beyond the elastic limit in which the elastic material has yielded all its elasticity permanently and has become plastic.

The Breaking Point

14

For a continued increase in load, a point is reached where the material cannot stand any further increase in load. At this point, the material breaks and this point is called the breaking point.

The Tensile stress, Tensile strain, and Young's Modulus

15

The tensile stress is the force exerted on an elastic material per unit cross-sectional area of it, that is:

$$\text{Tensile stress} = \frac{Force}{Cross - Sectional\ area} = \frac{F}{A} \quad \text{------------------------- (3)}$$

Its unit is $\frac{N}{m^2}$ which is same as Nm^{-2}.

The tensile strain is the extension produced on the elastic material per unit original length of it, that is:

$$\text{Tensile strain} = \frac{extension}{Original\ length} = \frac{e}{l} \quad \text{---------------------------------- (4)}$$

The tensile strain denotes the resulting effect of the applied force. It is a dimensionless quantity (has no unit) since it is a ratio of two quantities that have the same unit.

In terms of the tensile stress and the tensile strain, the Hooke's law for elastic bodies can be restated as follows: **provided the elastic limit is not exceeded, the tensile stress is proportional to the tensile strain.**

Mathematically, we write:
Tensile stress ∝ Tensile strain

$$\Rightarrow \frac{F}{A} \propto \frac{e}{l}$$

And $\frac{F}{A} = E\frac{e}{l}$ -- (5)

where E is a constant of the elastic material called the Young's modulus.

The Young's modulus is therefore defined as the ratio of the tensile stress to the tensile strain, that is:

$$\text{Young's modulus, } E = \frac{stress}{strain}$$

$$= \frac{F/A}{e/l}$$

$$= \frac{F}{A} \times \frac{l}{e}$$

Young's modulus, $E = \frac{Fl}{Ae}$ --(6)

We feel you should be able to attempt this question!

16

In an experiment to determine the Young's modulus for a wire, several loads are attached to the wire and the corresponding extensions measured. The tensile stress in each case depends on the
(A) load and the extension
(B) load and the radius of the wire
(C) radius of the wire and the extension
(D) extension and the original length of the wire

If you chose option B then you are very right!

17

From the definition of tensile stress, it depends on the force applied and on the cross sectional area of the elastic material.

The force applied is same as the load and the cross sectional area depends on the radius, so option B is the right answer.

Next, let's attempt this JAMB question!

18

If the stress on a wire is $10^7 Nm^{-2}$ and the wire is stretched from its original length of 10.00cm to 10.05cm. The young's modulus of the wire is

(A) $5 \times 10^4 Nm^{-2}$ (B) $5 \times 10^5 Nm^{-2}$ (C) $2 \times 10^8 Nm^{-2}$ (D) $2 \times 10^9 Nm^{-2}$

Solution!

19

Stress = $10^7 Nm^{-2}$
Original length = 10.00cm = 0.1m
Extension = (10.05 – 10.00) cm = 0.05cm = 0.0005 m

$$\Rightarrow \text{strain} = \frac{extension}{original length}$$

$$= \frac{0.0005m}{0.1m}$$

$$= 0.005$$

∴ Young's modulus = $\dfrac{stress}{strain}$

$= \dfrac{10^7 \, Nm^{-2}}{0.005}$

$= 2 \times 10^9 \, Nm^{-2}$

Another JAMB question for you to tackle!

20

The tendon in a man's leg is 0.01m long. If a force of 5N stretches the tendon by 2×10^{-5} m. Calculate the strain on the muscle.

(A) 5×10^6 (B) 5×10^2 (C) 2×10^{-3} (D) 2×10^{-7}

Did you get the right answer? Find out!

21

Strain = $\dfrac{extension}{original \; length}$

We are given:
Extension e = 2×10^{-5} m
Original length = 0.01m

∴ Strain = $\dfrac{2 \times 10^{-5}}{0.01}$

$= 2 \times 10^{-3}$

So option C is correct!

Notice that there was no need of the 5N force given in the question. This was, perhaps, JAMB's way of testing if you know what parameters are required to solve the problem, so bravo if you got it!

Work done in stretching or compressing elastic materials

22

The work done in stretching or even compressing an elastic material is given by the product of the average force and the displacement (or extension, e, produced).

If we assume that the force applied on the material is gradually increased from 0 to a final value F, then we compute the average force as:

Average force = $\dfrac{\text{initial force} + \text{final force}}{2}$ = $\dfrac{0+F}{2}$ = $\dfrac{F}{2}$

Therefore, the work done = $\dfrac{F}{2} \times e = \dfrac{1}{2} Fe$ ---------------------------- (7)

But from Hooke's law; F = ke

$\Rightarrow$ work done, w = $\dfrac{1}{2} Fe$ = $\dfrac{1}{2} ke^2$ -- (8)

The work done in stretching or compressing an elastic material is stored as elastic potential energy in the material.

For graphical purpose!

23

If we gradually increase the force on an elastic material from 0 to a value F, then the extension will also increase from 0 to a value (let's say, e).

Then we make a Force-Extension curve for the material as shown below.

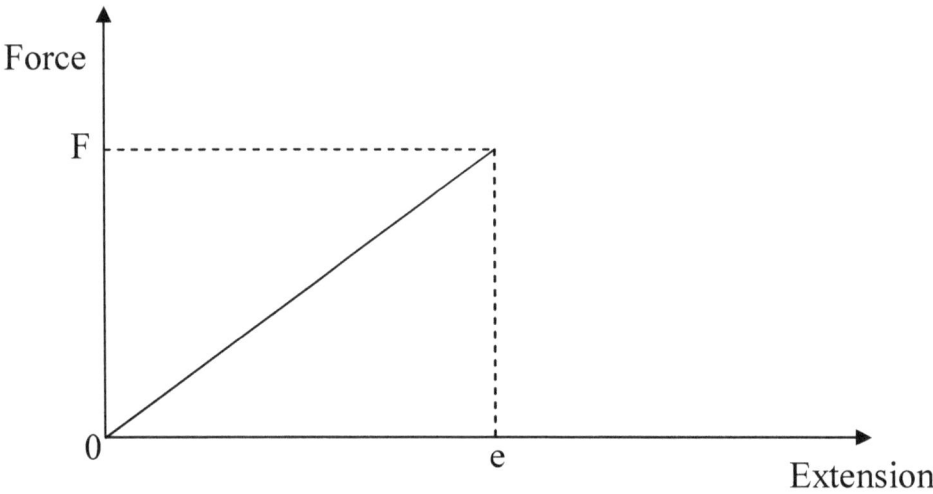

The energy stored in the material in this case is **the area under the graph**
That is the area of the triangle $= \frac{1}{2} base \times height$

$$= \frac{1}{2} e \times F$$

$$= \frac{1}{2} Fe$$

which is same as equation (7) above.

In general, the area under the Force-Extension curve of an elastic material represents the energy stored in it.

Let's take more JAMB Questions

24

A spring of force constant $1500 Nm^{-1}$ is acted upon by a constant force of 75N. Calculate the potential energy stored in the spring.

(A) 1.9J (B) 3.2J (C) 3.0J (D) 5J

Solution!

25

The potential energy stored in the spring is given as:

$$w = \frac{1}{2}Fe \quad \text{--- (vii)}$$

But from Hooke's law; $e = \frac{F}{k}$

Substituting this into equation (vii) gives

$$w = \frac{1}{2}F \times \frac{F}{k}$$

$$= \frac{F^2}{2k}$$

$$= \frac{75^2}{2 \times 1500}$$

$$= 1.9 \text{ J}$$

Here is another JAMB question, just for you!

26

A spring of length 25cm is extended to 30cm by a load of 150N attached to one of its ends. What is the energy stored in the spring?

(A) 3750J　　　　(B) 2500J　　　　(C) 3.75J　　　　(D) 250J

Did you get 3.75J? Then you are right!

27

Energy stored in the spring is
$$w = \frac{1}{2}Fe$$

where we are given: F = 150N
and e = 30cm − 25cm = 5cm = 0.05m

$$\therefore w = \frac{1}{2} \times 150N \times 0.05m$$

= 3.75Nm which is the same as 3.75J

Yet another graphical JAMB question!

28

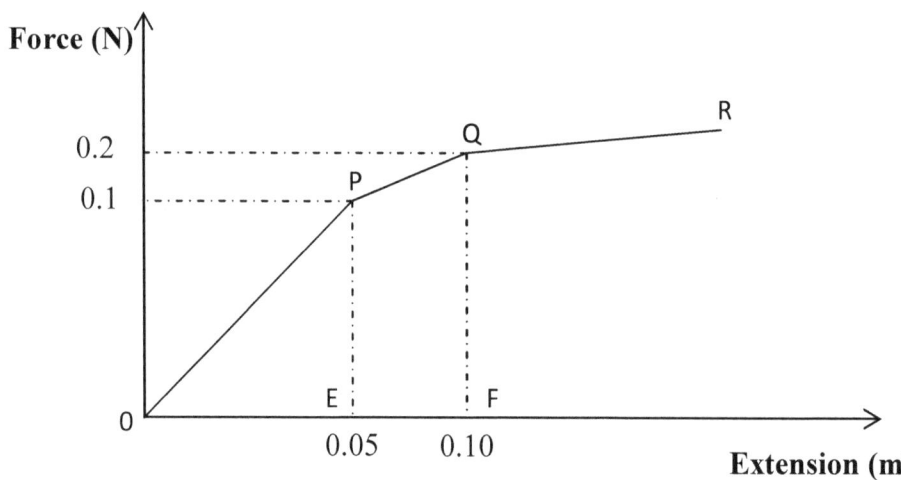

The diagram above shows the Force-Extension curve of a piece of wire. The energy stored when the wire is stretched from E to F is

(A) 7.5×10^{-3} J   (B) 2.5×10^{-3} J   (C) 1.5×10^{-2} J   (D) 7.5×10^{-1} J

Solution!

29

Recall the energy stored in an elastic material is given by the area under the Force-Extension curve.

For this question, we are only interested in the region of the curve from E to F; the energy stored when the wire is stretched from E to F is the area under this portion of the graph. That is the area of the trapezium EPQF.

Area of the trapezium EPQF $= \frac{1}{2}(a + b) \times h$

where a and b are lengths of the parallel sides, and h is the distance between them.

$$= \frac{1}{2}(PE + QF) \times EF$$

$$= \frac{1}{2}(0.1 + 0.2) \times (0.10 - 0.05)$$

$$= \frac{1}{2}(0.3) \times (0.05)$$

$$= 7.5 \times 10^{-3} J$$

Beautiful if you understand it! And that brings us to the end.
Next, a couple of exercises for you to play with!

Exercises

30

1. If a force of 50N stretches a wire from 20m to 20.01m, what is the amount of force required to stretch the same material from 20m to 20.05m?

(A) 250N   (B) 2000N   (C) 100N   (D) 50N

2.

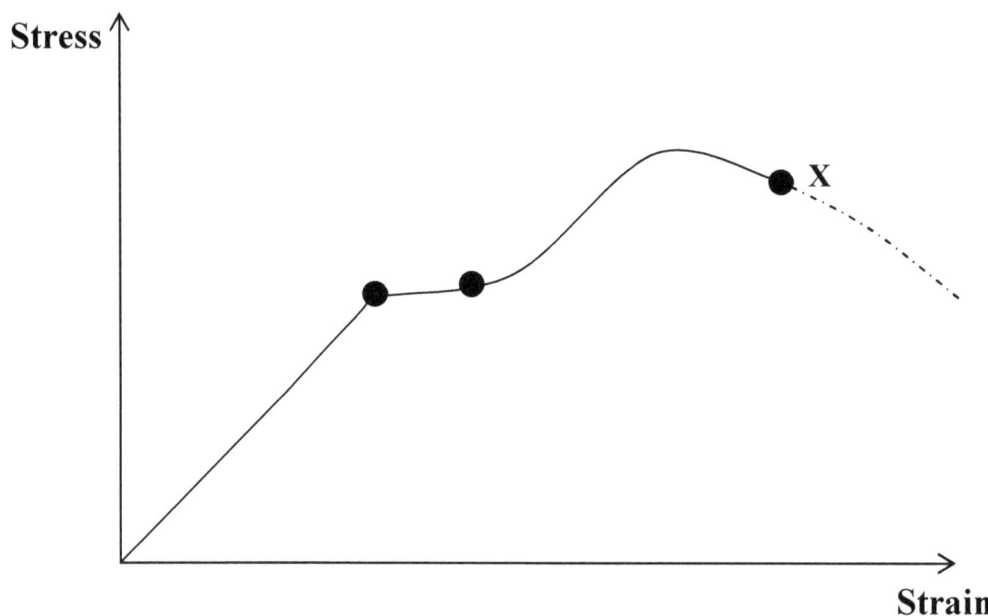

The diagram above represents the stress/strain graph for a typical metal rod. What does X represent?

(A) Breaking point   (B) Yield point   (C) Elastic limit   (D) Proportional limit

3. A load of 20N on a wire of cross-sectional area $8\times10^{-7}m^2$ produces an extension of 10^{-4}m. Calculate Young's modulus for the material of the wire if its length is 3m.

(A) $7.0\times10^{11}Nm^{-2}$ (B) $7.5\times10^{11}Nm^{-2}$ (C) $8.5\times10^{11}Nm^{-2}$ (D) $9.0\times10^{11}Nm^{-2}$

4. The energy contained in a wire when it is extended by 0.02m by a force of 500N is

(A) 5J (B) 10J (C) 10^3J (D) 10^4J

5. A force of 40N is applied to a wire 4m long and produces an extension of 0.24mm. If the diameter of the wire is 2.00mm, calculate the:
(i) stress on the wire
(ii) strain in the wire
[Take $\pi=3.142$]

6. A spiral spring of natural length 20.0cm has a scale pan hanging freely in its lower end. When an object of mass 40g is placed in the pan, its length becomes 21.80cm. If the first mass is removed and another object of mass 60g is placed in the pan, the length becomes 22.05cm. Calculate the mass of the scale pan.
[Take g = $10ms^{-2}$]

7. A group of students performed a spring experiment and obtained the following results:

load (g)	0	20	40	60	80	100	120	140
length (mm)	50	58	70	74	82	90	98	106
extension (mm)								

(a) What is the length of the spring when un-stretched?
(b) Copy and complete the table.
(c) Plot a graph of the data **(Plot x = load, y = spring extension)**.
(d) One of the results is wrong, which?
(e) What is the force constant of the spring?

(f) What load would give an extension of 30 mm?

(g) What would be the spring length for a load of 50g?

8. The diagram below represents the graph of the force applied in stretching a spiral spring against the corresponding extension produced within its elastic limit.

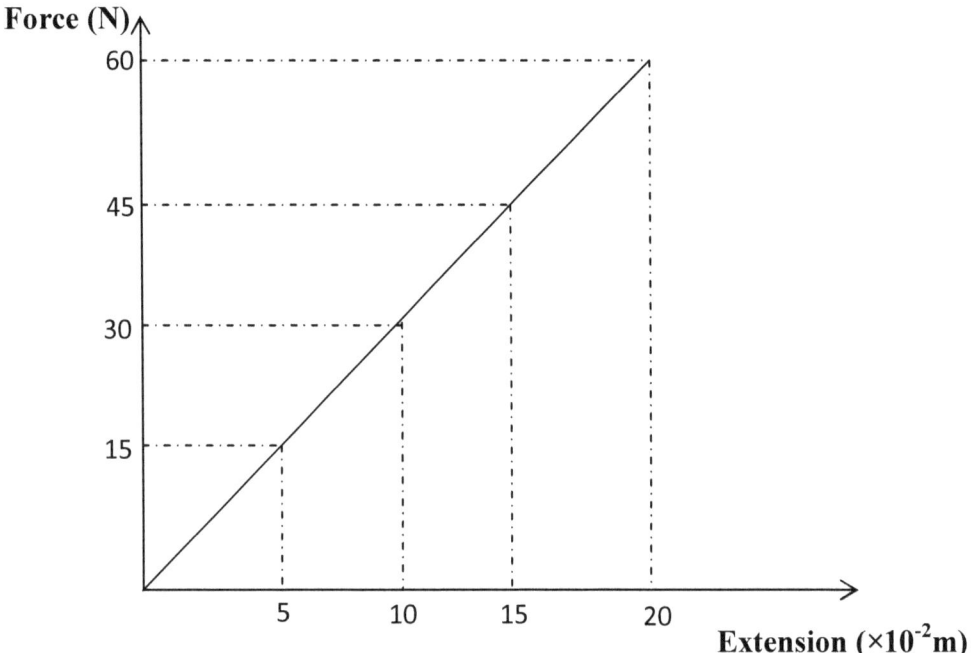

Using the notations on the graph, determine the:
   (a) force constant of the spring
   (b) work done in stretching the spring from 10×10^{-2} m to 20×10^{-2} m

Solution to exercises

31

1. A
2. A
3. B
4. A
5. (i) $12.73 \times 10^6 \, Nm^{-2}$ (ii) 6.0×10^{-5}
6. 104g

7. (a) 50mm

[The length of the spring when un-stretched is the length when there is zero load on it]

(b)

load (g)	0	20	40	60	80	100	120	140
length (mm)	50	58	70	74	82	90	98	106
extension (mm)	0	8	20	24	32	40	48	56

(c)

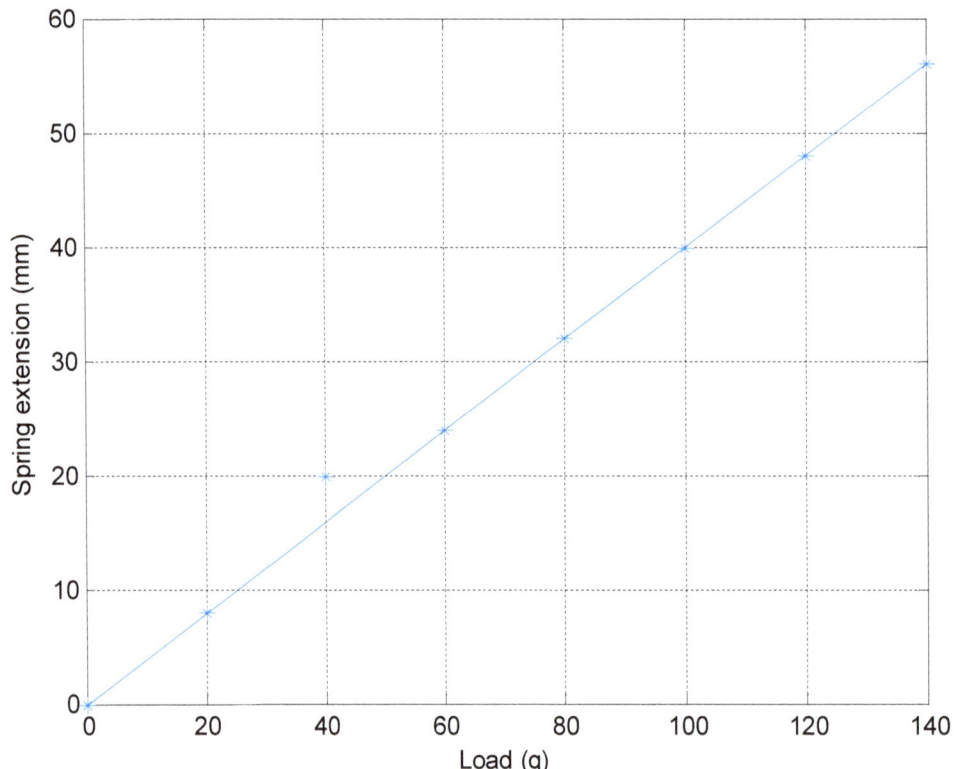

(d) The result corresponding to when the load is 40g is wrong. That is the result which is out of the line in the graph above.

(e) 2.5g/mm   [That is the reciprocal of the slope of the graph above. Alternatively one can use Hooke's law; F=ke. Using any pair of readings on the table (except the wrong one), we can substitute for the load and extension to get the force constant k].

(f) 75g
(g) 70mm

8. (a) 300Nm^{-1} (b) 4.5J

www.ingramcontent.com/pod-product-compliance
Lightning Source LLC
Chambersburg PA
CBHW050435180526
45159CB00006B/2543